Table of Contents
Math Homework Booklet
Kindergarten

one	2-5
two	6-9
three	10-13
one-three	14-15
four	16-19
five	20-23
six	24-27
one-six	28-31
seven	32-35
eight	36-39
nine	40-43
ten	44-47
one-ten	48-54
sequencing	55-57
more/less	58-59
circle	60
square	61
triangle	62
rectangle	63
oval	64
diamond	65
heart	66
star	67
calendars	68-69
time	70-72

Numero Uno

___ the picture.

___ the number 1.

___ the word one.

one

✏️ each set of 1.

⭕ one.

Where Are the Ones?

○ all the 1's in the picture.

○ the picture.

How many chimps? _____

the spaces: 1 = blue
• = gray
one = red

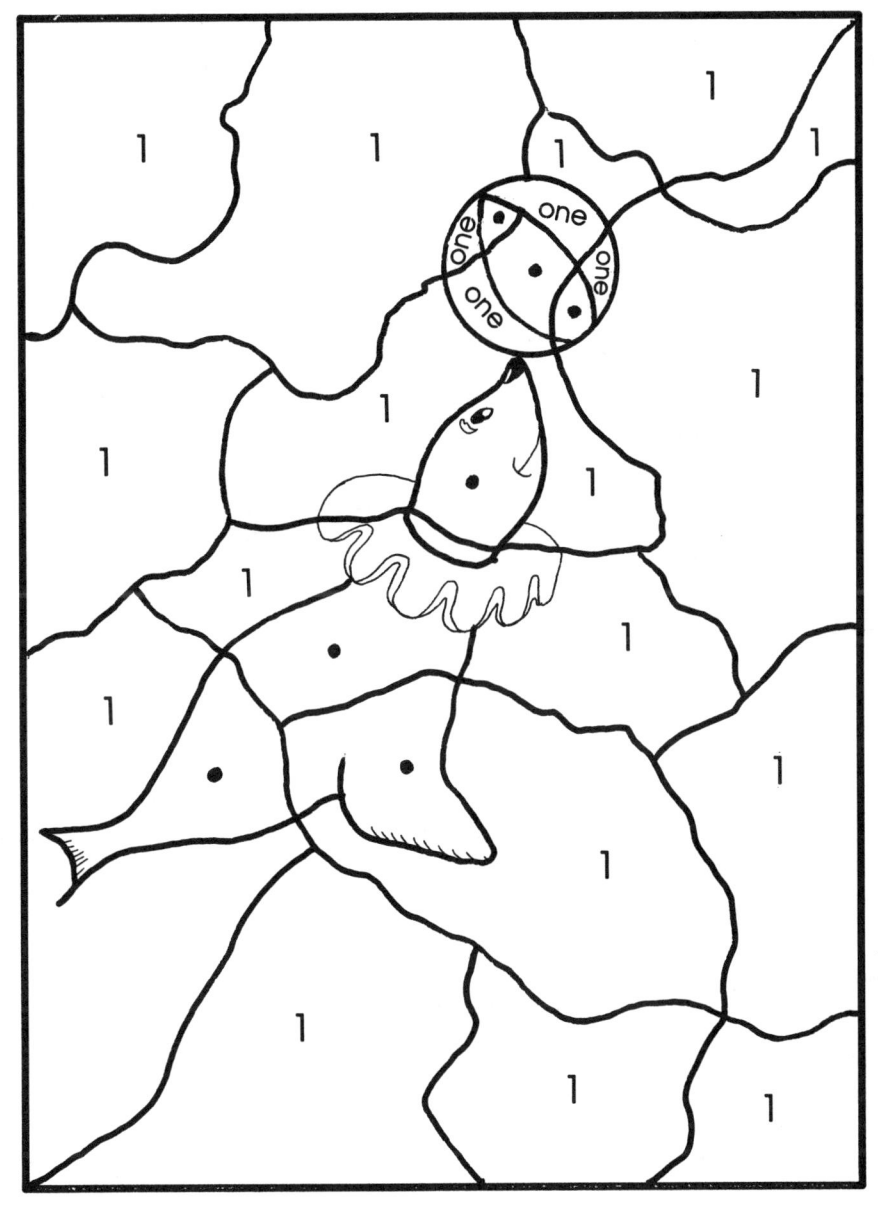

Terrific Twos

 the picture.

 the number 2.

2

 the word two.

two

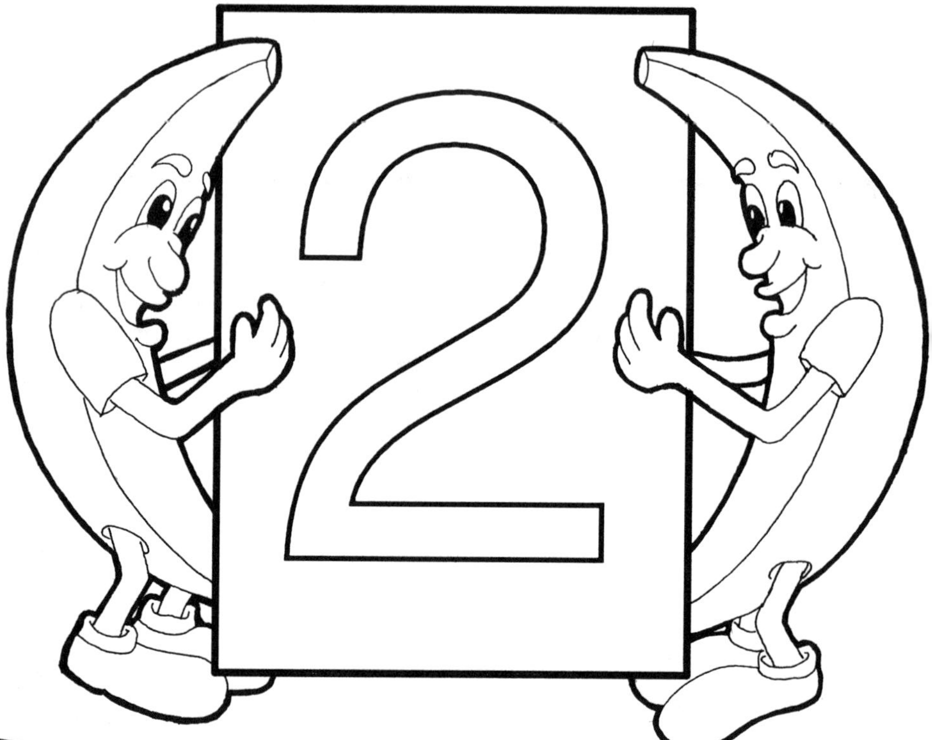

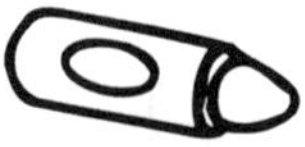

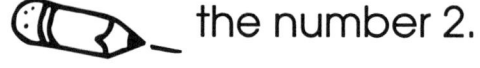

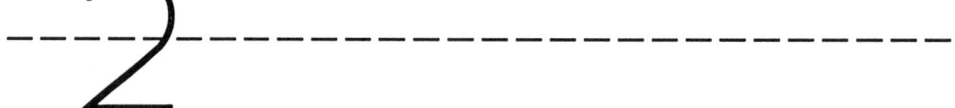

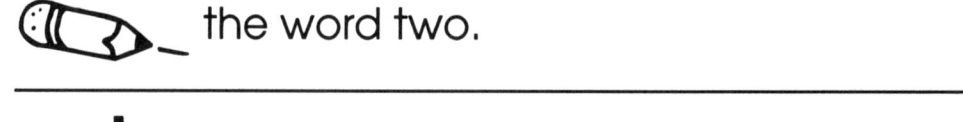

two

 each set of 2.

 two.

Hide and Seek

✏️ all the 2's in the picture.

🖍️ the picture.

How many apples? _____

the spaces: two = blue   2 = red
•• = green   1 = black

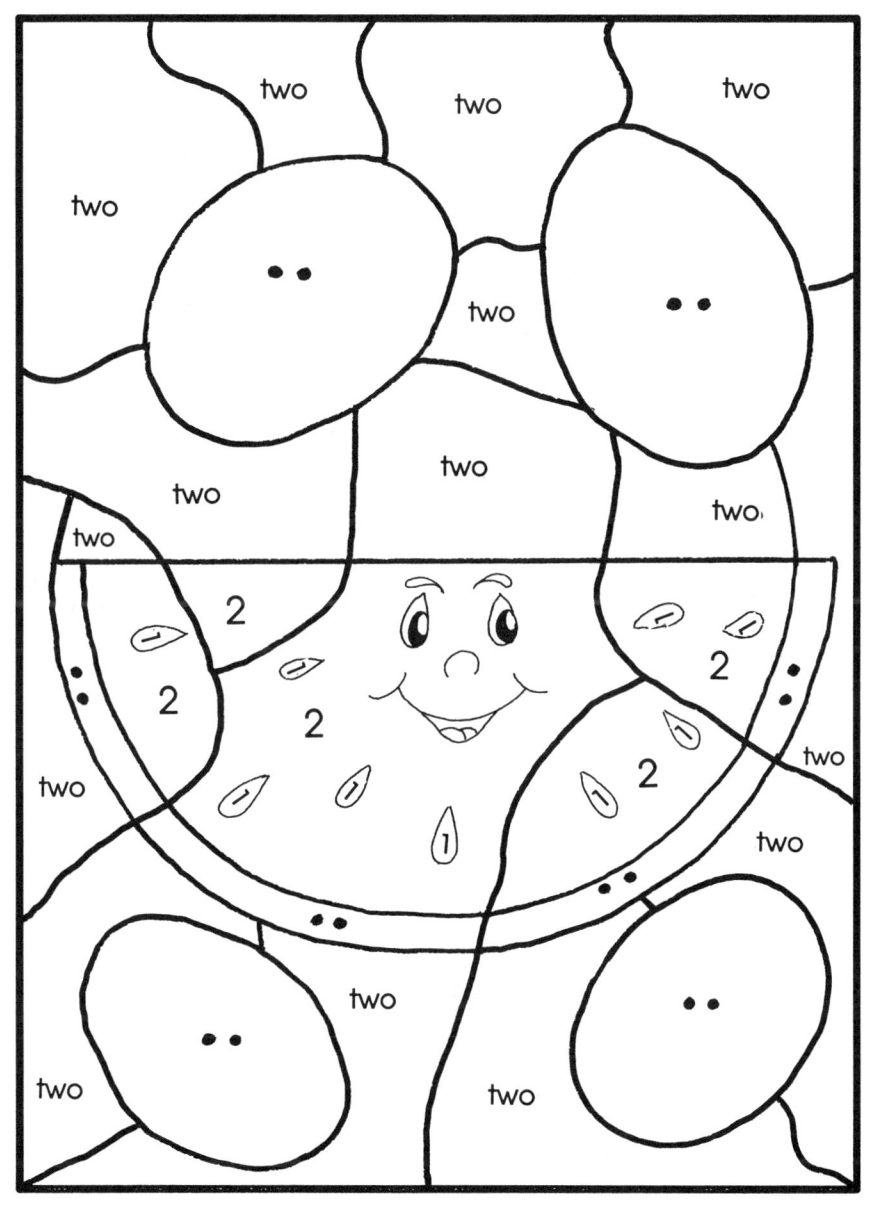

Tremendous Threes

Color the picture.

Trace the number 3.

3

Trace the word three.

three

three

✏️ each set of 3.

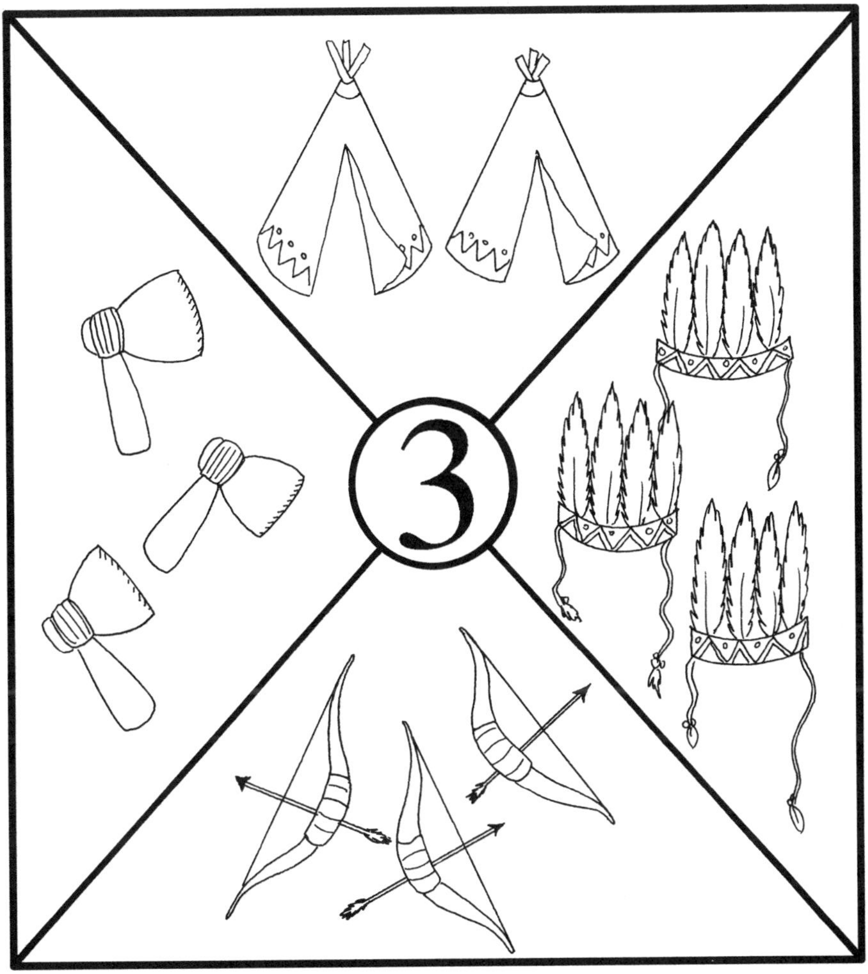

✏️ ⭕ three.

Let's Find the Threes!

🖍️ all the 3's in the picture.

🖍️ the picture.

How many Indians? _____

three

the spaces: 3 = red   one = yellow
              ⋰ = gray   2 = brown

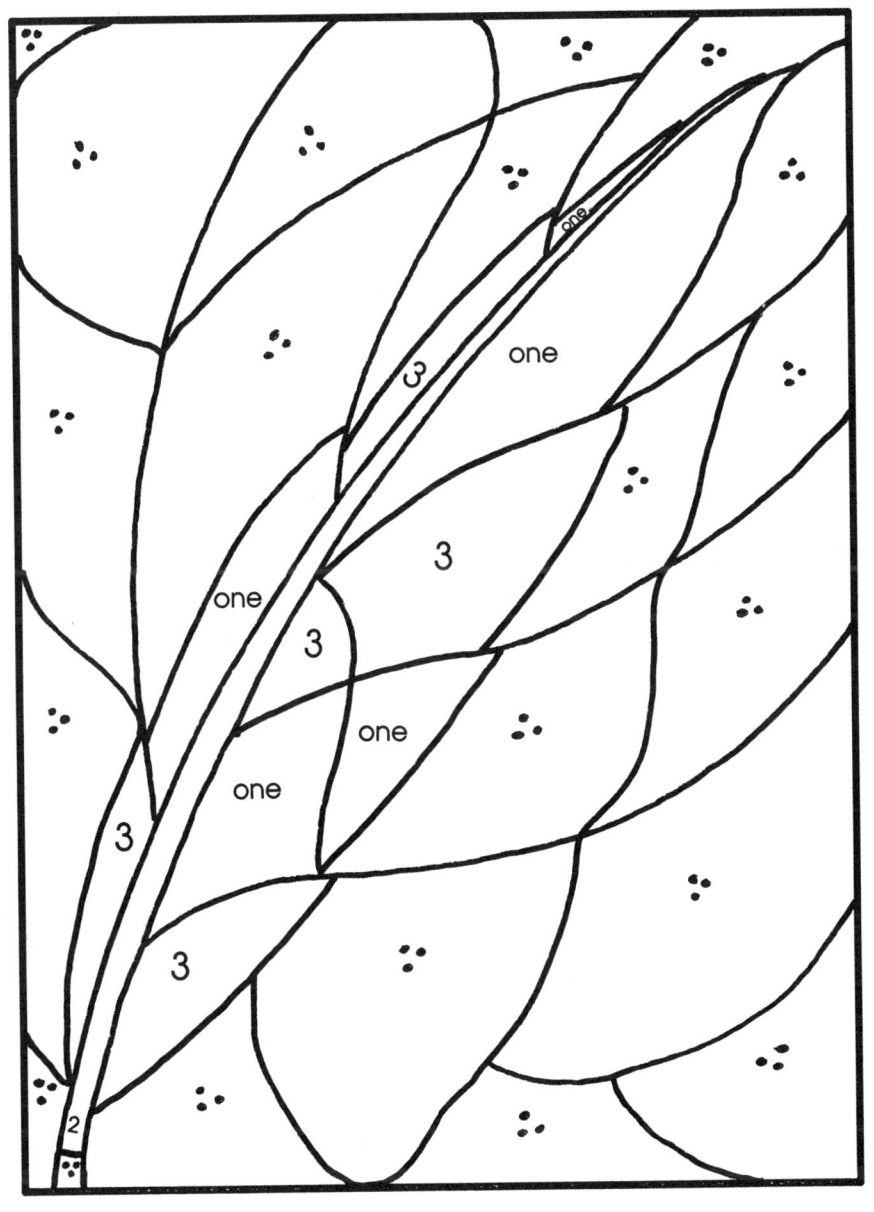

Match 'Em Up!

🖍 __ the words.

🖍 __ a line to match the numbers, words and pictures.

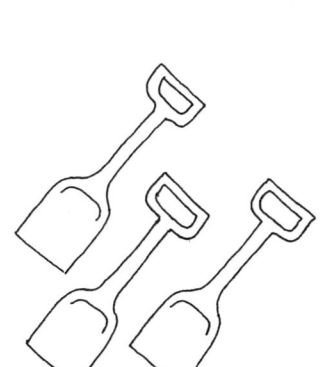

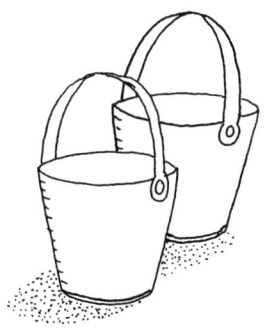

one-three

How many?

The Funny Four

🖍 the picture.

✏️ the number 4.

✏️ the word four.

four

◯ each set of 4.

◯ four.

Circus Fun

◯ all the 4's in the picture.

◯ the picture.

How many balloons? _____

four

the spaces: four = green   :: = red
4 = yellow   2 = purple
two = orange

High Five

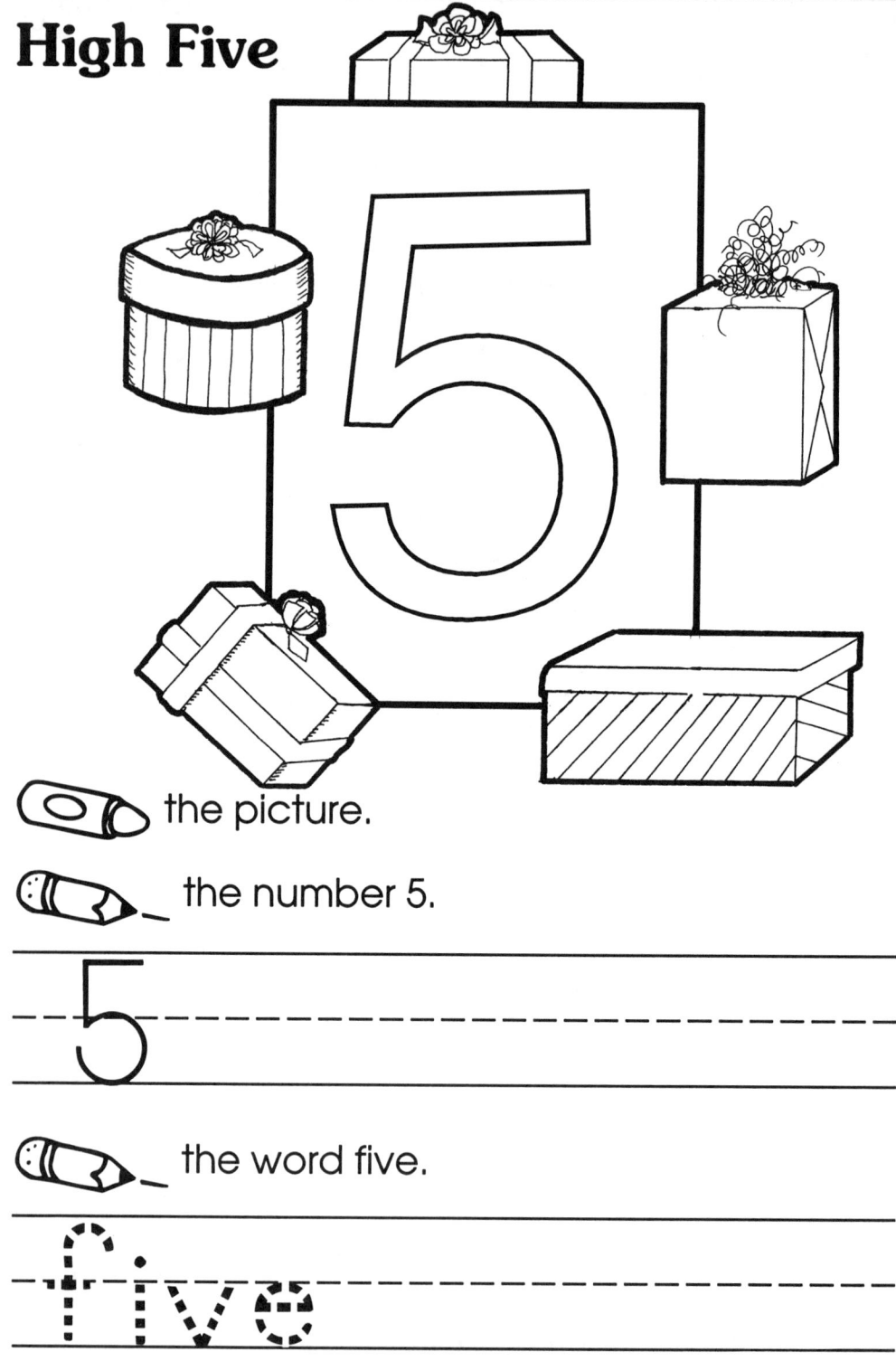

Color the picture.

Trace the number 5.

Trace the word five.

five

🖍 each set of 5.

✏️ five.

Birthday Magic

✏️ all the 5's in the picture.

🖍️ the picture.

How many balloons? _____

five

the spaces: 5 = yellow   ⋮ = brown
             five = pink 4 = blue
             2 = green   three = purple

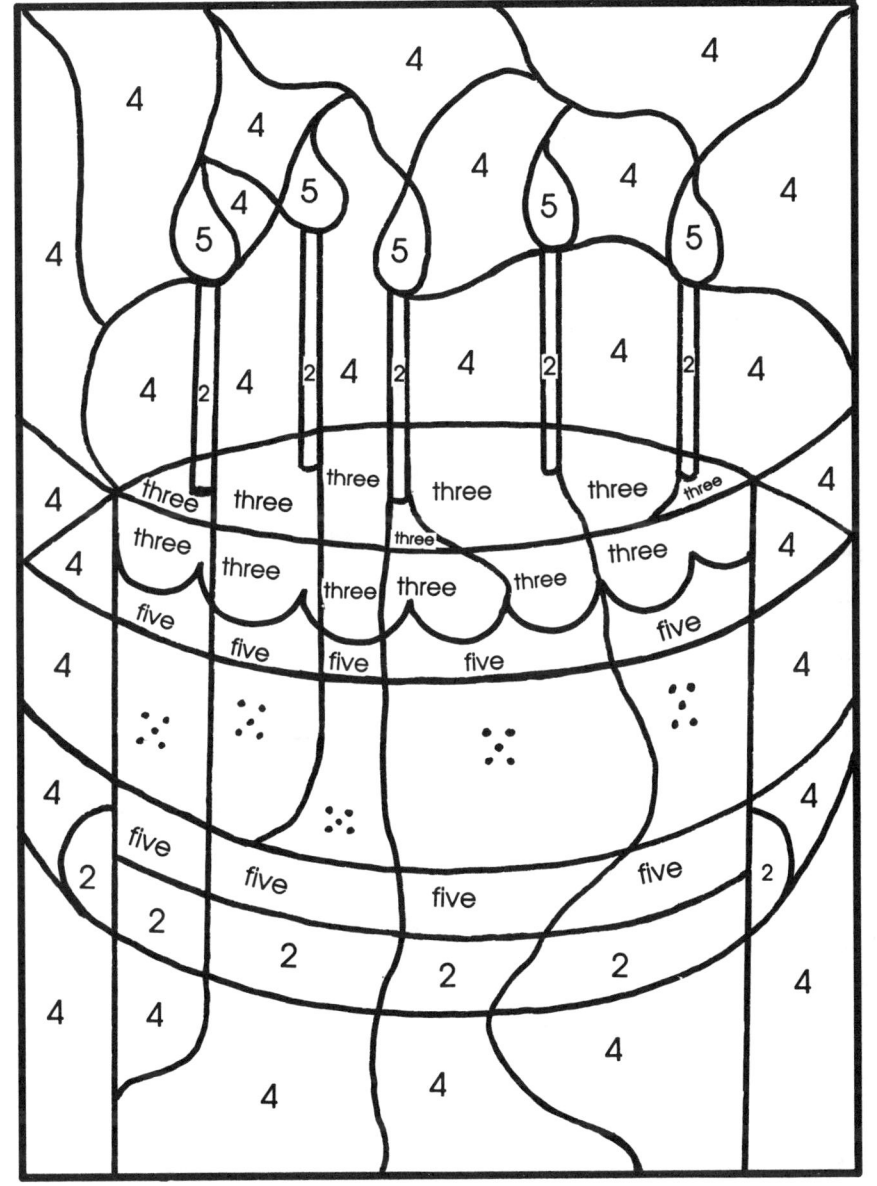

Super Six

 the picture.

 the number 6.

6

 the word six.

six

six

 each set of 6.

 six.

Farmer in the Dell

✏️ all the 6's in the picture.

🖍️ the picture.

How many chicks? _____

six

the spaces: 6 = blue ⋮⋮ = yellow
six = brown

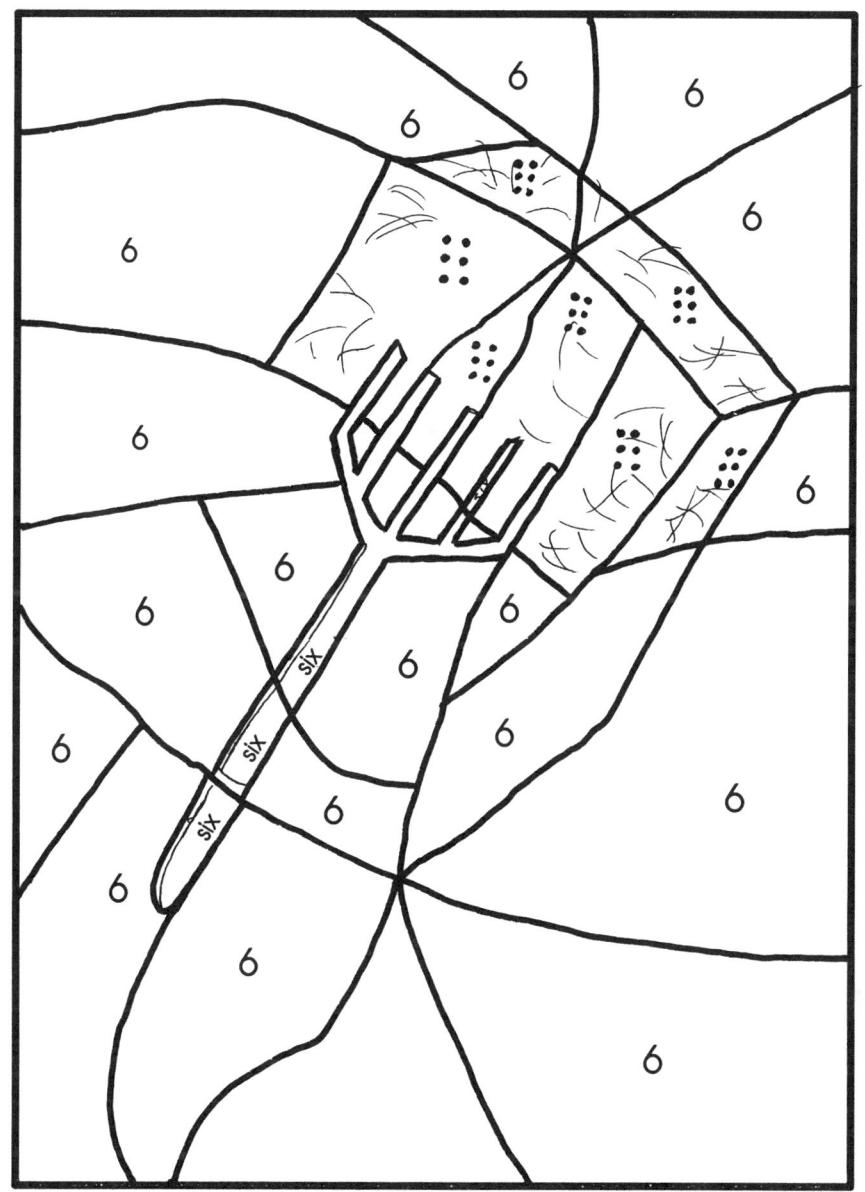

Picnics Are Fun!

___ a line from 1-6 on each set of numbers.

___ the picture.

one-six

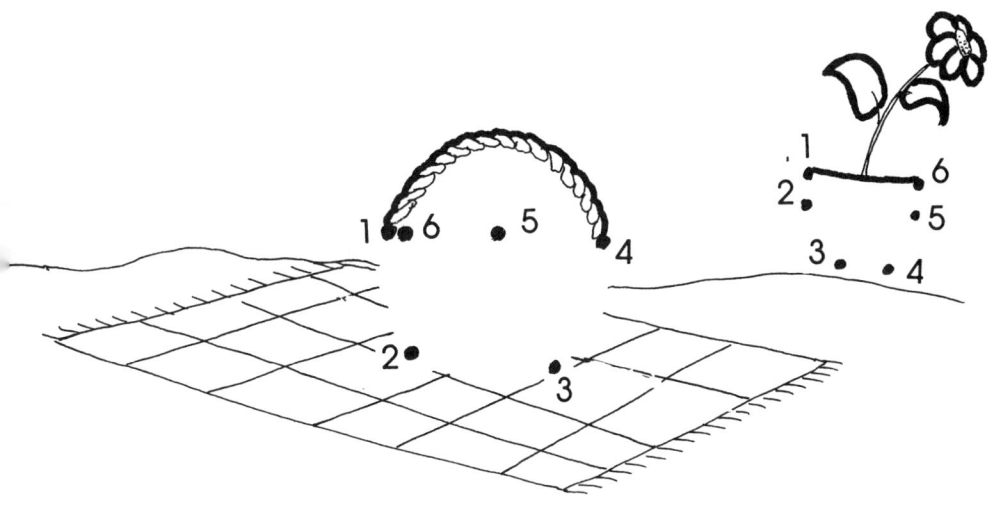

Lots of Money

 the words in the puzzle. Look ↓ and →.

one
two
three
four
five
six

one-six

 the number of items in each row.

Silly Seven

◯ the picture.

✏ the number 7.

✏ the word seven.

Leap Frog

◯ all the 7's in the picture.

◯ the picture.

How many lily pads? _____

 the spaces: 7 = blue   ⋮⋮ = green   seven

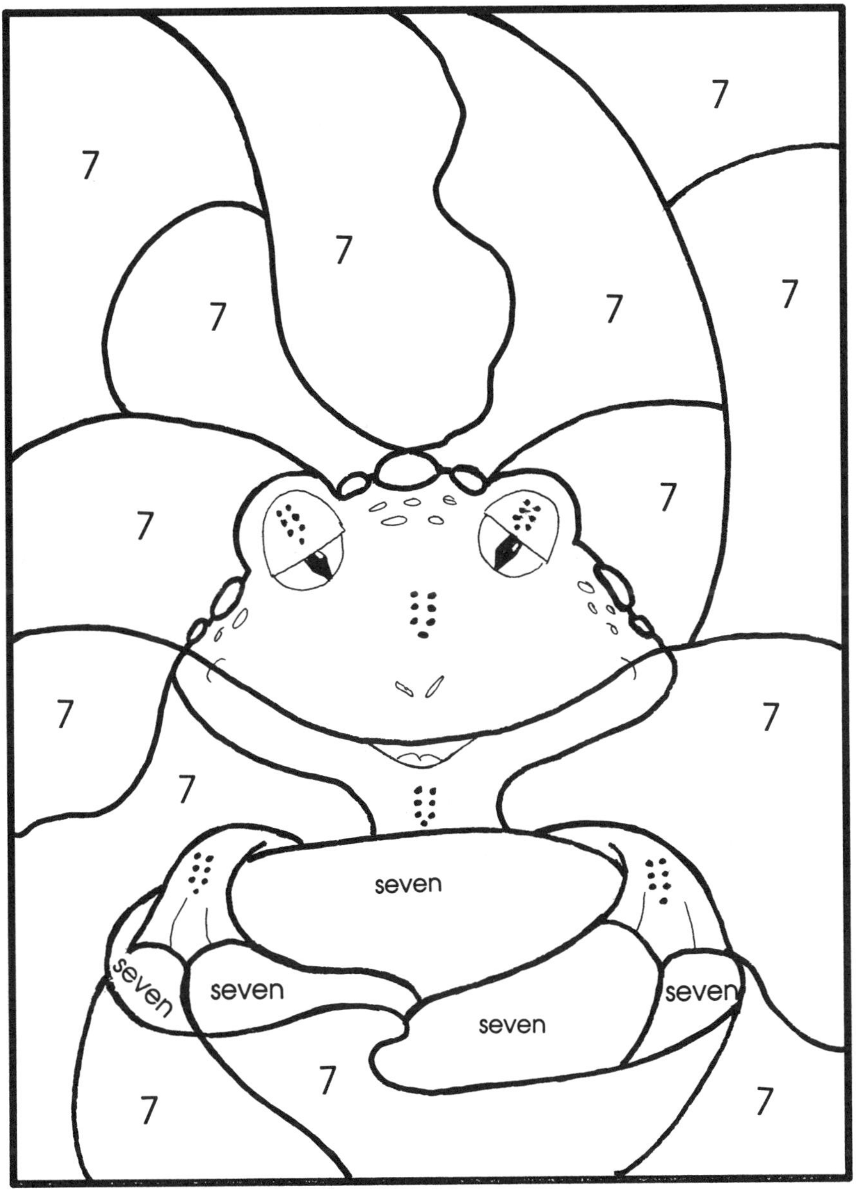

seven = brown

Magic Eight

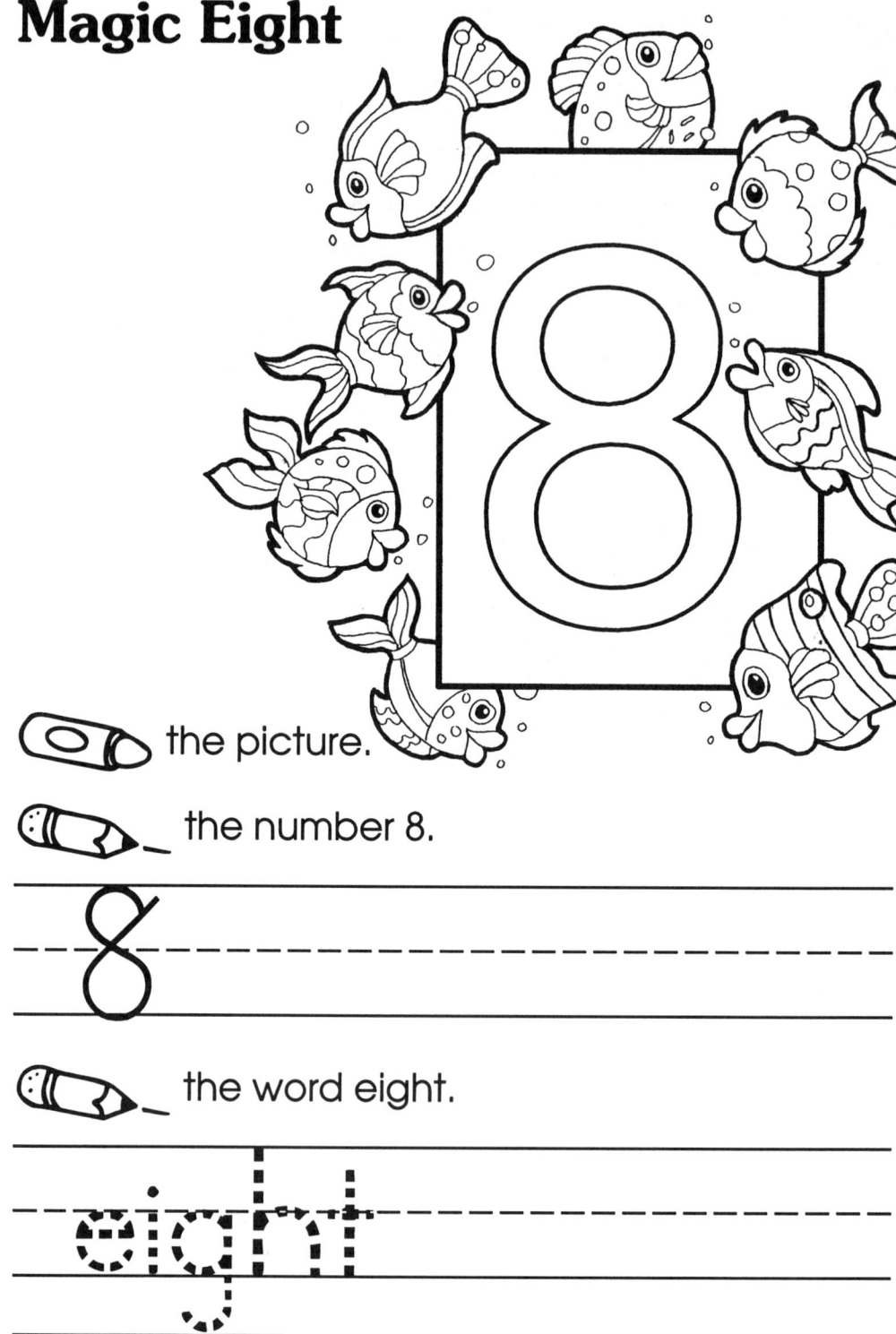

 the picture.

 the number 8.

8

 the word eight.

eight

Tutor's Guide

This tutor's guide contains answer keys for Kindergarten Math. Pull it out from the book to use as a guide.

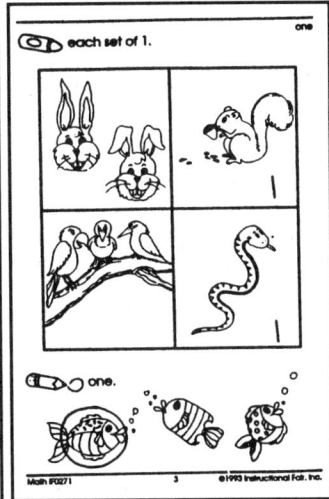

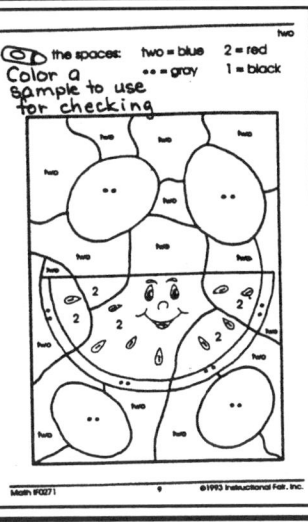

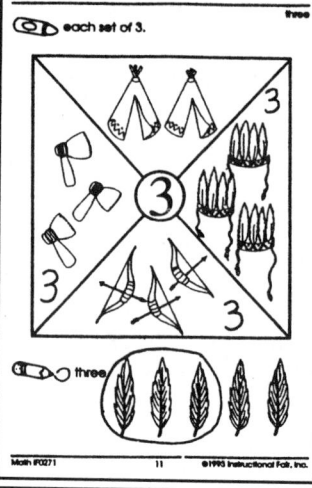

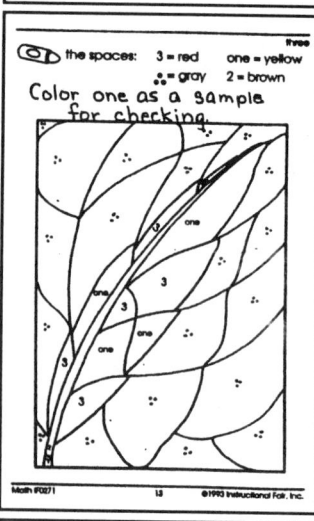

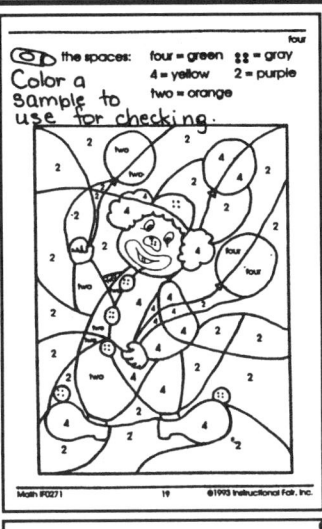

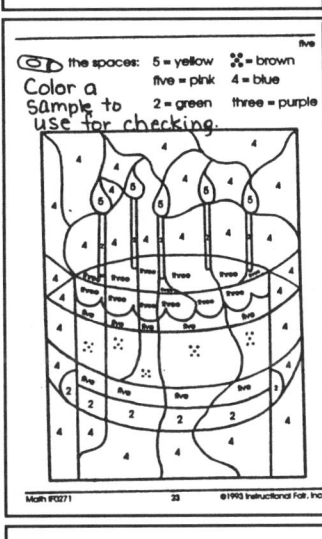

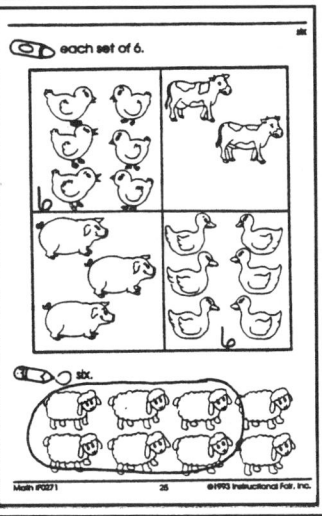

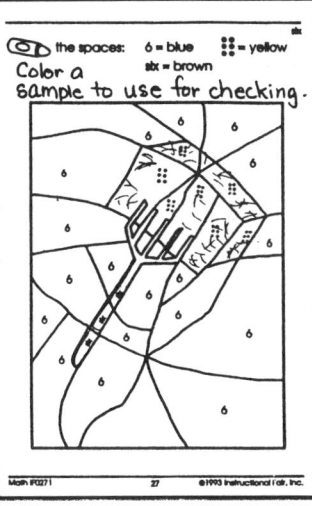

Tutor's Guide IF0271   C   ©1993 Instructional Fair, Inc.

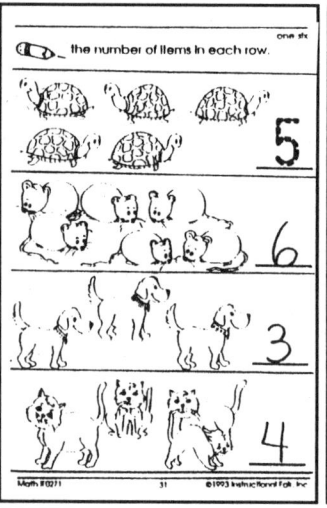

Tutor's Guide IF0271 — D — ©1993 Instructional Fair, Inc.

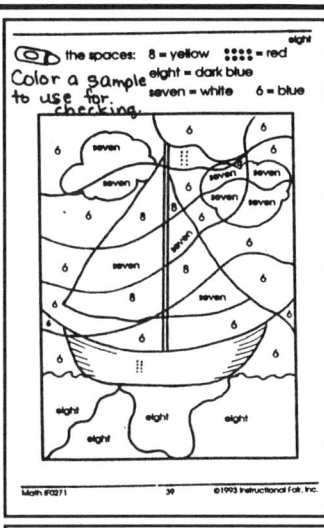

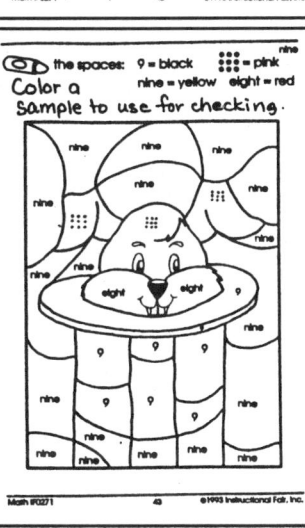
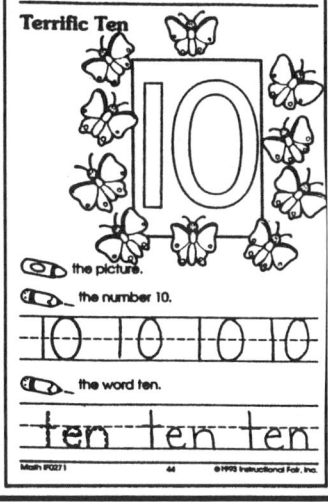

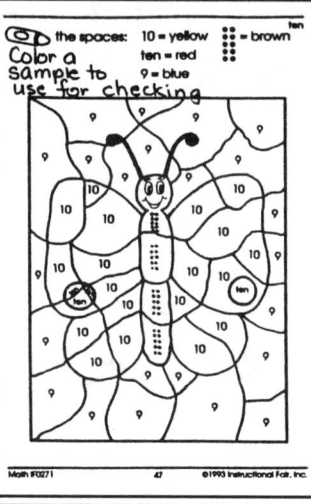

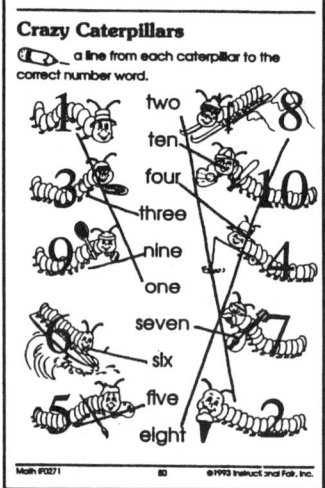
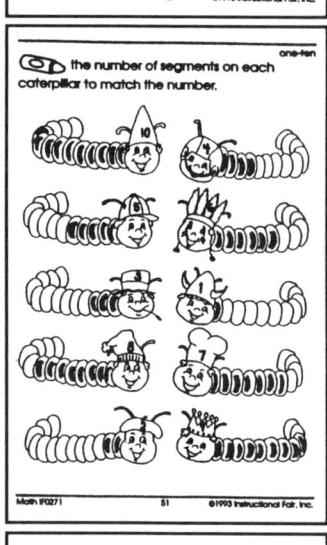
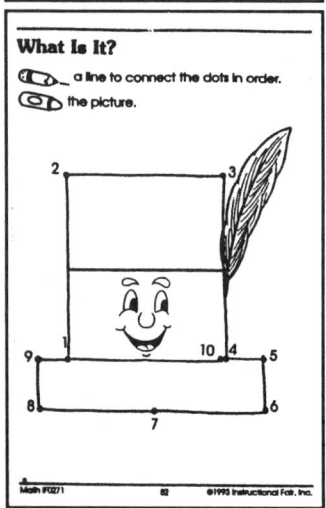

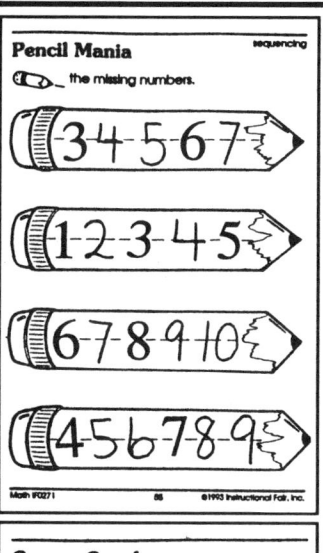

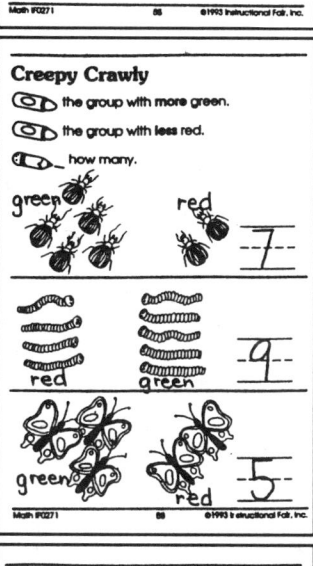

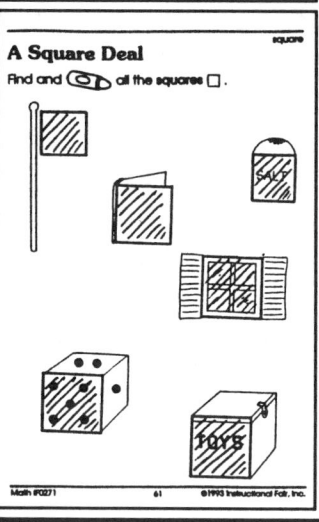

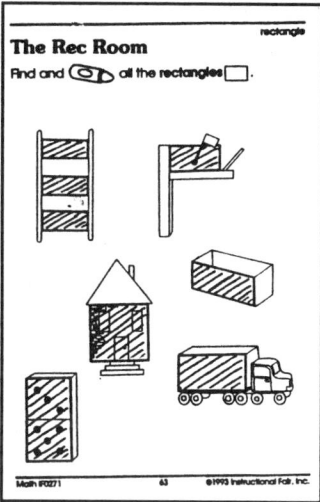

Ogling Ovals
Find and 🖍 all the ovals ○.

Diamonds Are Forever
Find and 🖍 all the diamonds ◊.

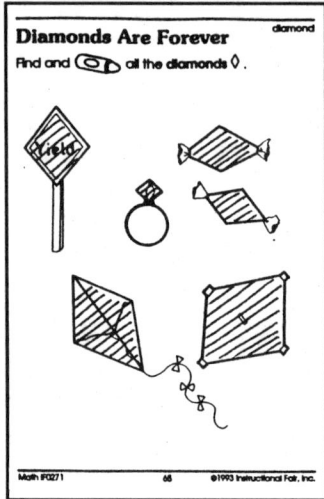

Heart Throb
Find and 🖍 all the hearts.

Starry Night
Find and 🖍 all the stars ☆.

What Day Is It?
🖍 In the missing numbers on the calendar. Answer the questions below.

Sunday	Monday	Tuesday	Wednesday	Thursday	Friday	Saturday
	1	2	3	4	5	6
7	8	9	10	11	12	13
14	15	16	17	18	19	20
21	22	23	24	25	26	27
28	29	30	31			

How many Mondays do you see? 5
How many Fridays do you see? 4
How many Sundays do you see? 4
How many Wednesdays do you see? 5

🖍 a line to match the calendar with the correct season.

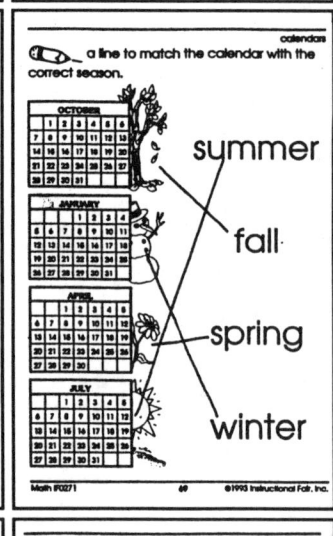

summer

fall

spring

winter

What Time Is It?
🖍 the time each clock shows.

1 o'clock

= 8 o'clock

= 2 o'clock

= 3 o'clock

🖍 the time each clock shows.

= 5 o'clock

= 6 o'clock

= 8 o'clock

= 10 o'clock

= 12 o'clock

Happy Face
🖍 the missing numbers on the clock below.

eight

🖍 and 🖍 8 fish.
🖍 and 🖍 8 castles.
🖍 8 snails.

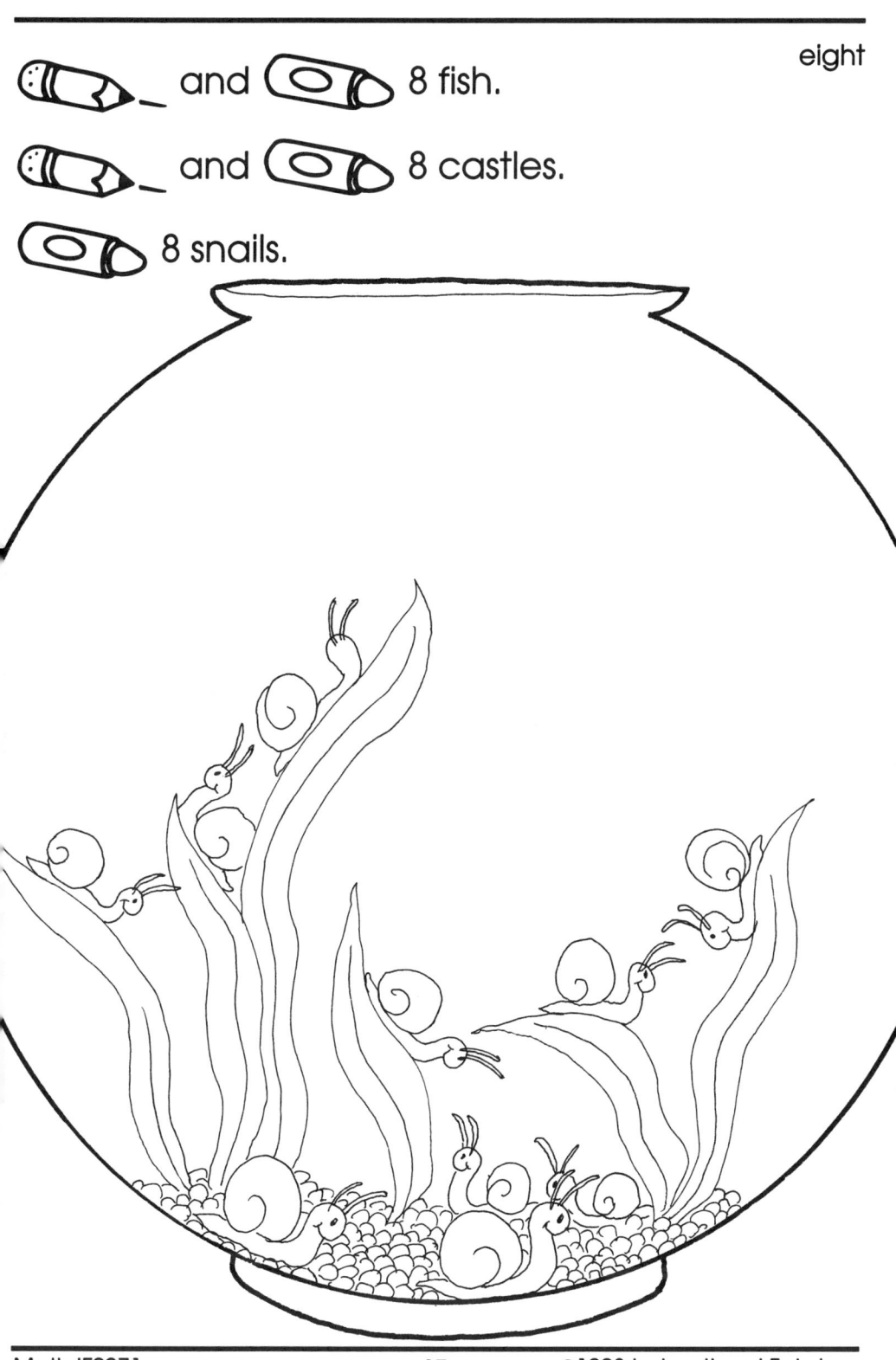

How Many 8's Are There?

◯ all the 8's in the picture.

the picture.

How many fish? _____

the spaces: 8 = yellow   ⁜ = red   eight = dark blue   seven = white   6 = blue

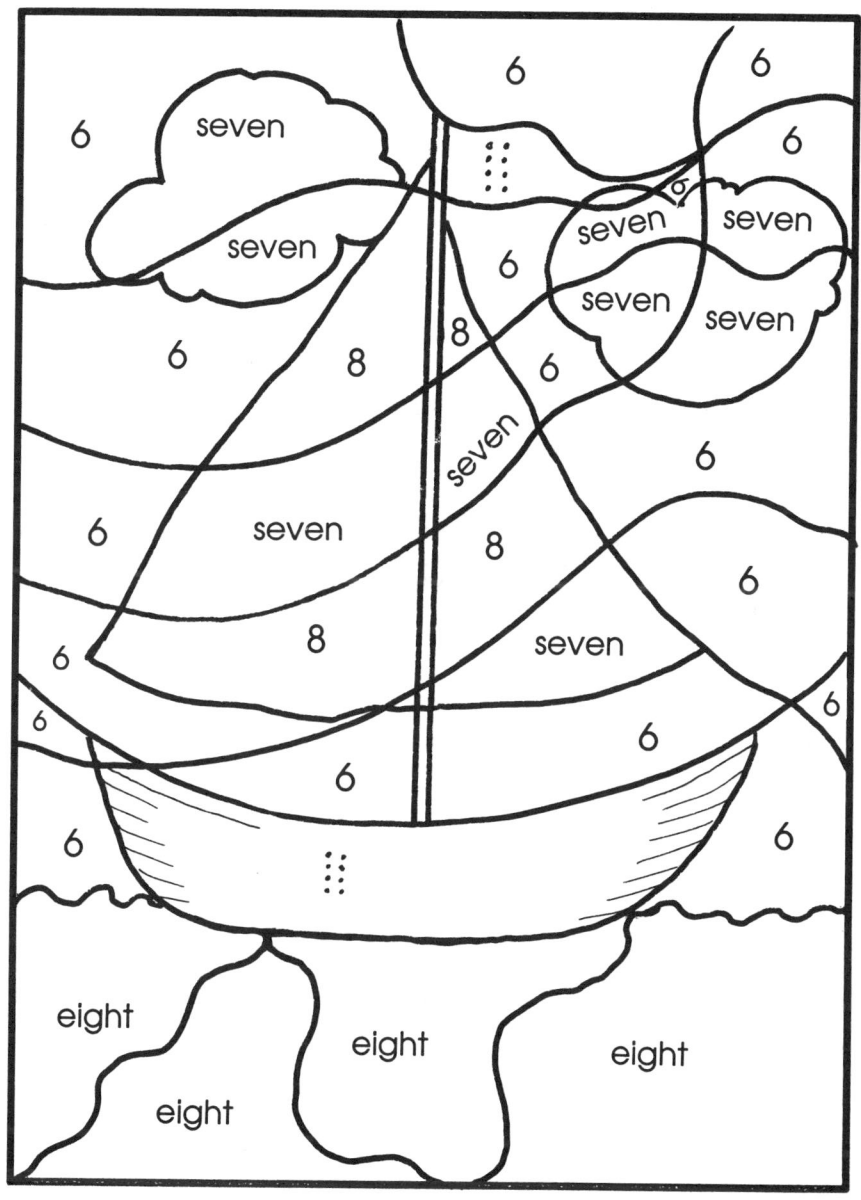

Nifty Nine

🖍 the picture.

✏️ the number 9.

9

✏️ the word nine.

nine

Hop! Hop! Hop!

🖊 ⭕ all the 9's in the picture.

🖍 the picture.

How many vegetables? _____

the spaces: 9 = black   ⋮⋮⋮ = pink   nine
nine = yellow   eight = red

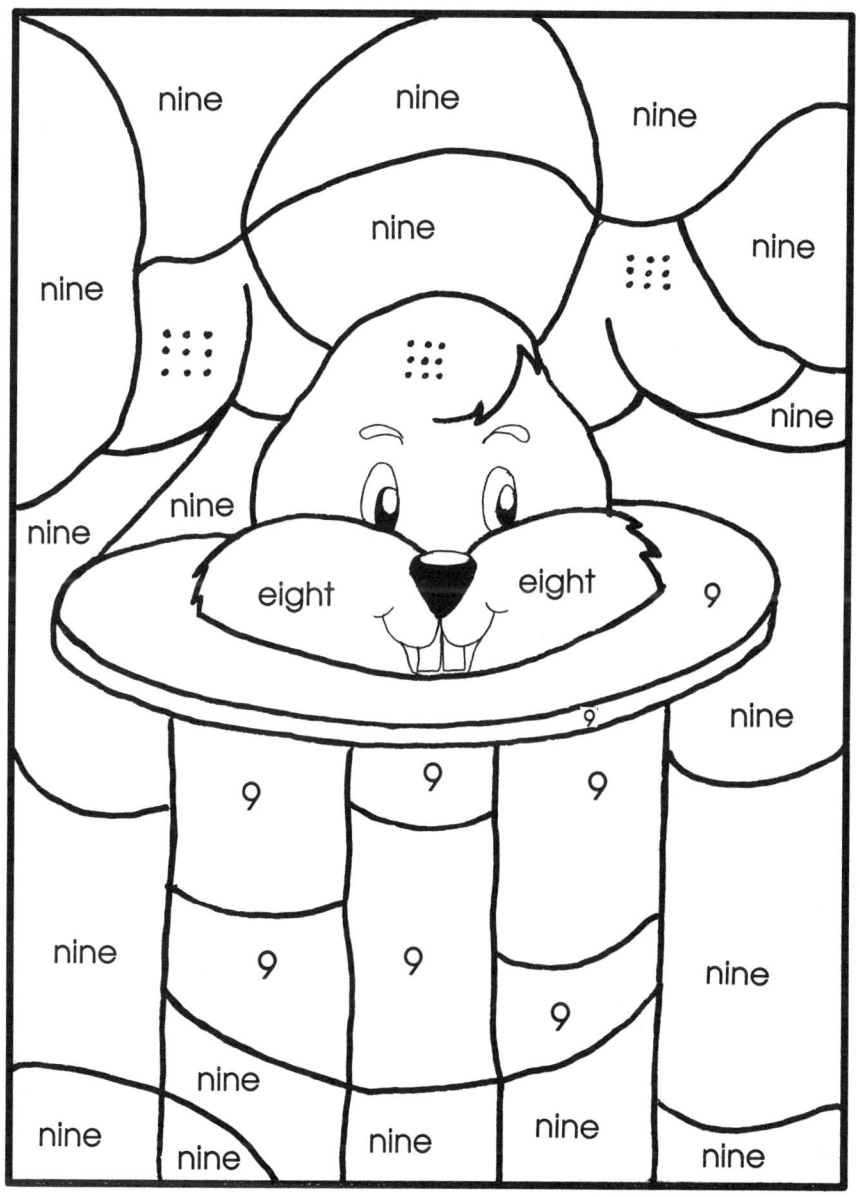

Terrific Ten

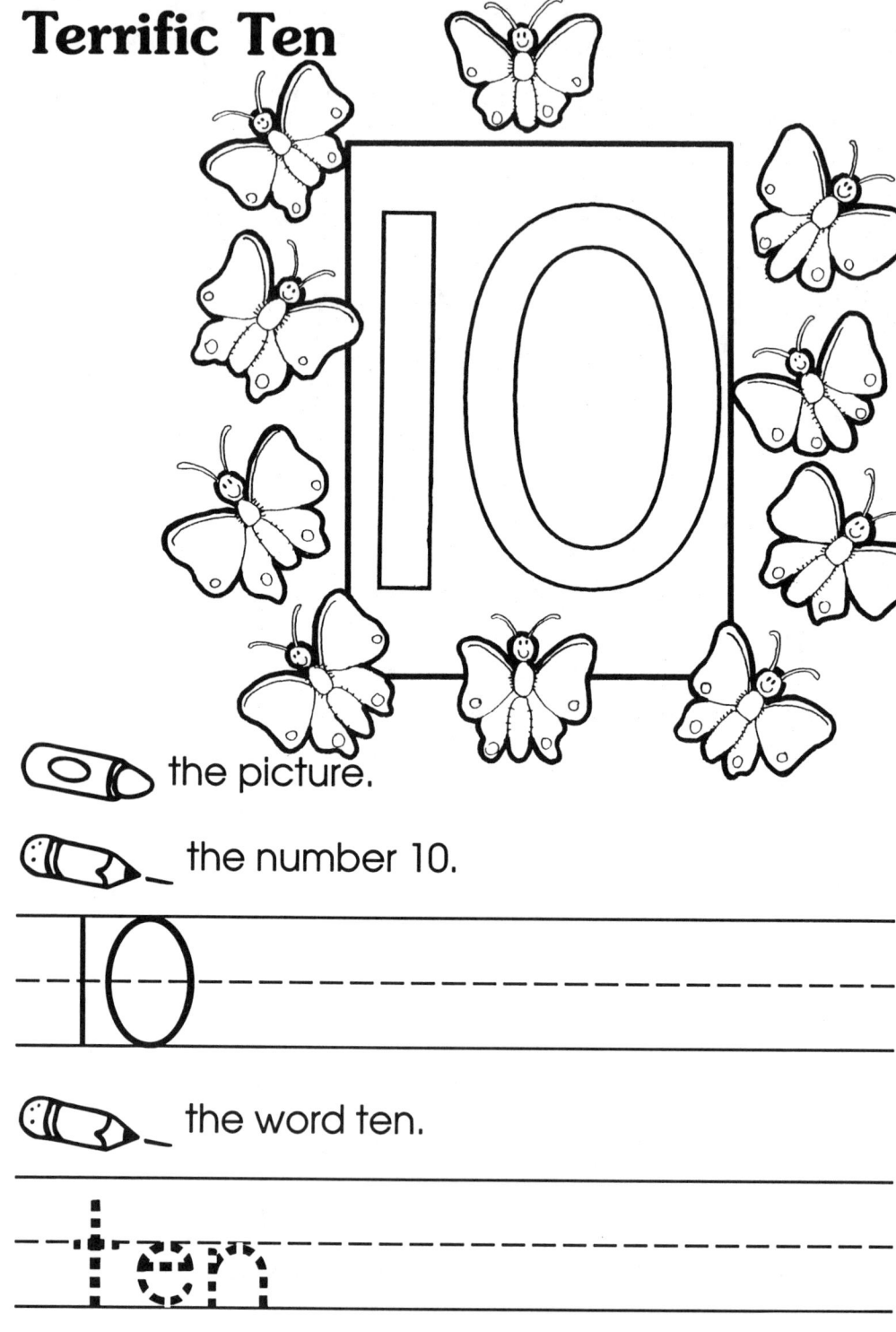

 the picture.

 the number 10.

 the word ten.

ten

🖍 10 clouds.

🖍 10 flowers.

🖍 and 🖍 10 spots on the butterfly.

The Countryside

◯ all the 10's in the picture.

▭ the picture.

How many flowers? _____

the spaces: 10 = yellow ⋮⋮ = brown ten
ten = red
9 = blue

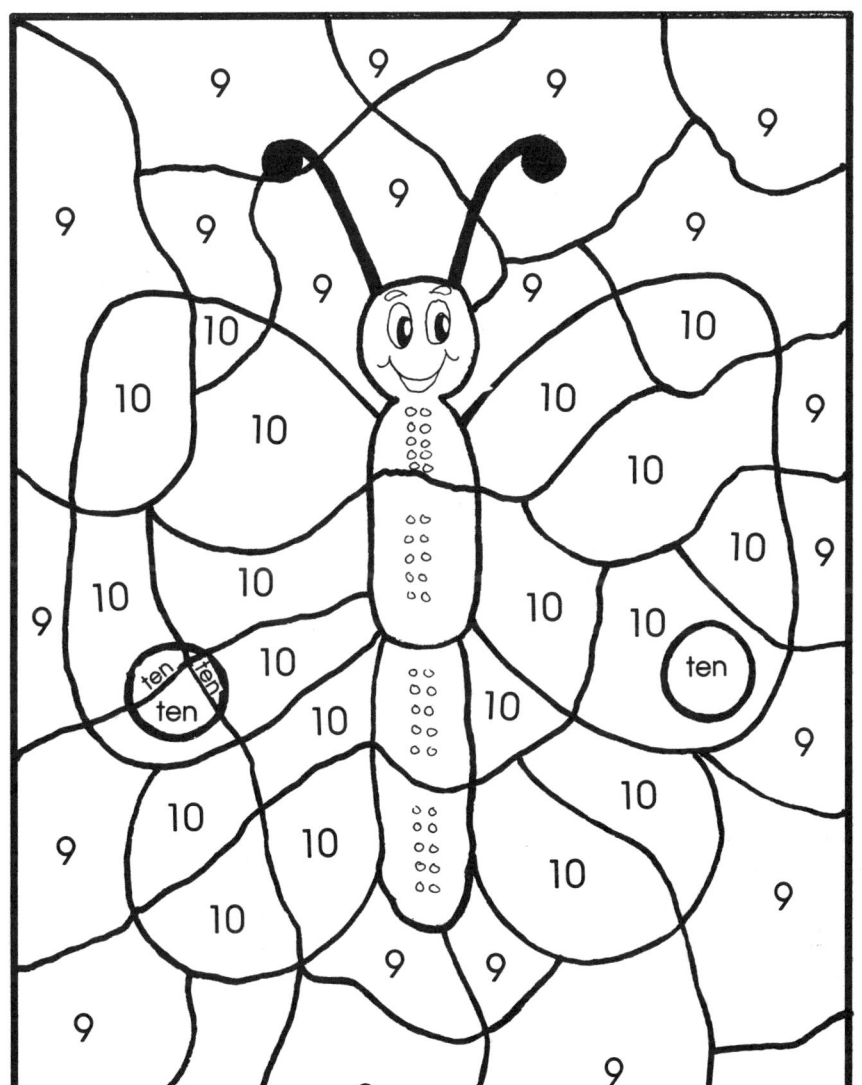

The Air Show

How many?

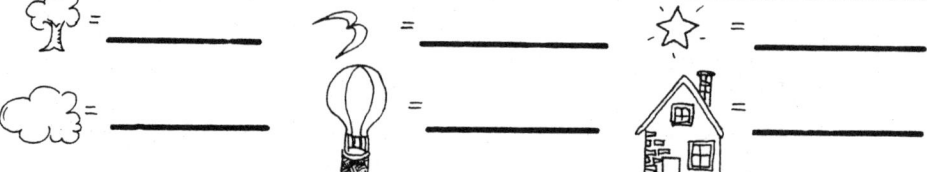

one-ten

Crazy Caterpillars

✏️ a line from each caterpillar to the correct number word.

two
ten
four
three
nine
one
seven
six
five
eight

one-ten

🖍 the number of segments on each caterpillar to match the number.

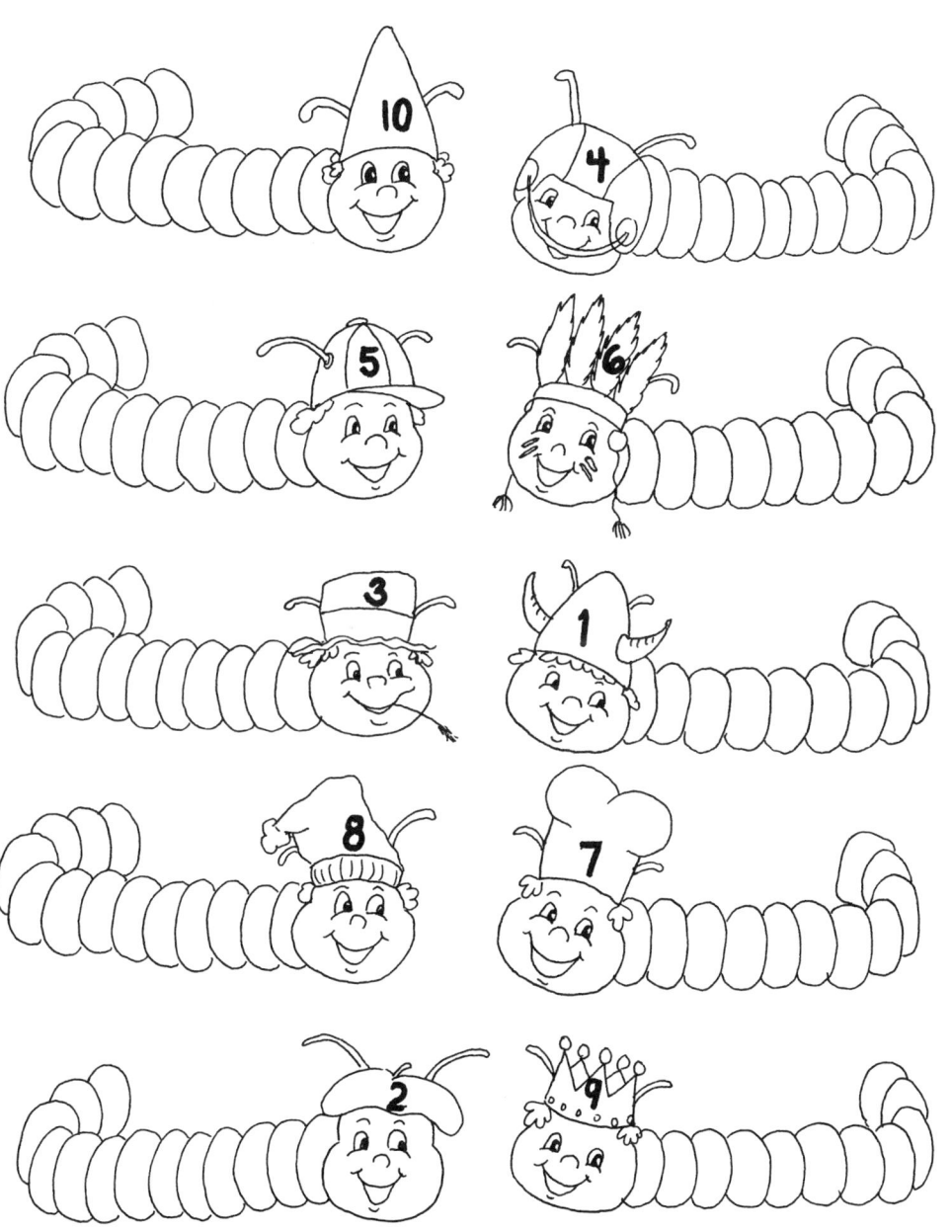

What Is It?

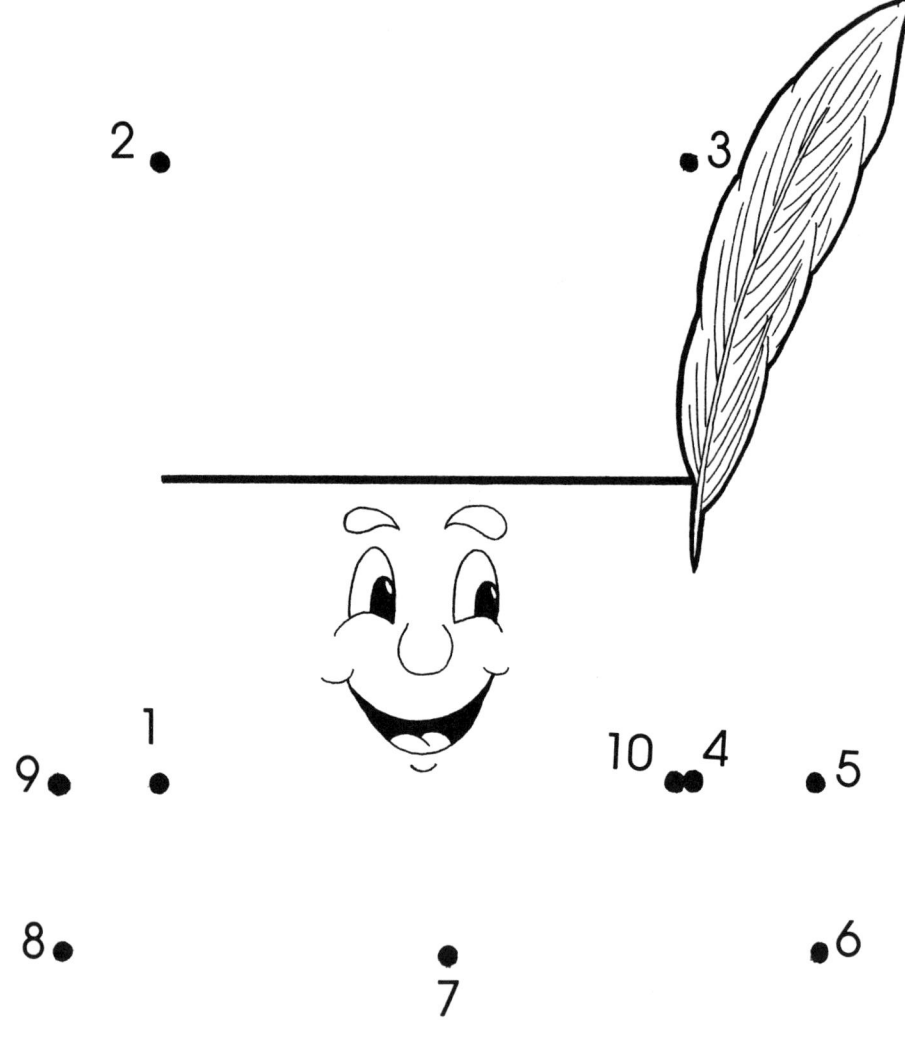 a line to connect the dots in order.
the picture.

one-ten

 the spaces:

one = purple   four = orange   seven = red
two = yellow   five = blue   eight = pink
three = black   six = green   nine = brown

Hats! Hats! Hats!

one-ten

 the number to match the word.

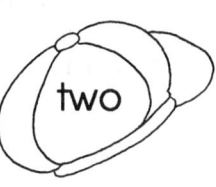
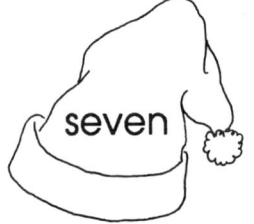
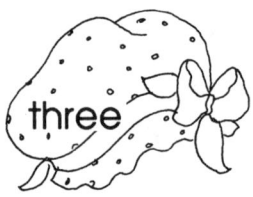

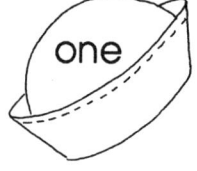
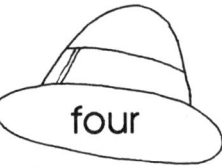

Pencil Mania

sequencing

✏️ the missing numbers.

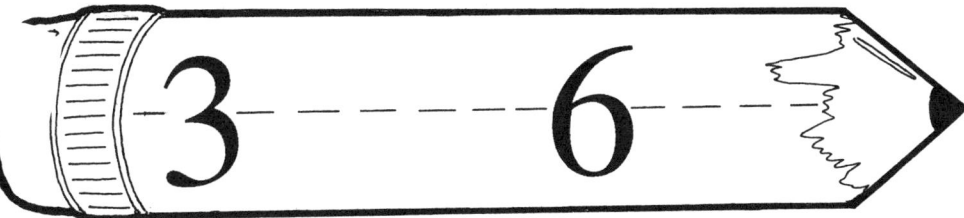

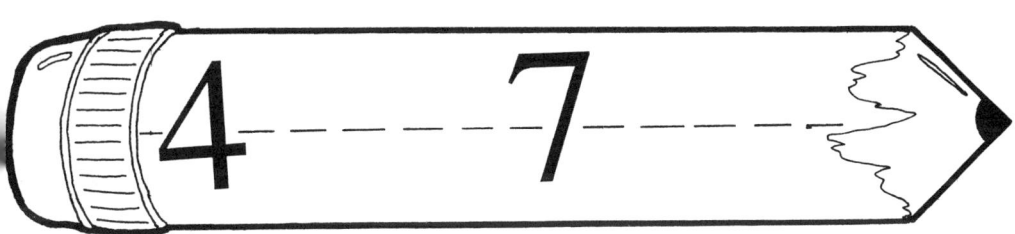

What Comes First?

 the number that comes before.

_____ 3   _____ 8

_____ 10   _____ 9

 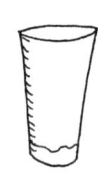   

_____ 2   _____ 5

sequencing

 the number that comes after.

1 _____   0 _____

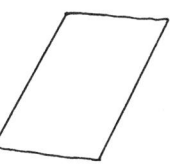 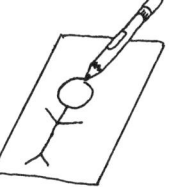   

5 _____   7 _____

9 _____   3 _____

Creepy Crawly

🖍 the group with **more** green.

🖍 the group with **less** red.

✏ how many.

more/less

⊙▷ the group with **more** green.

⊙▷ the group with **less** red.

✎ how many.

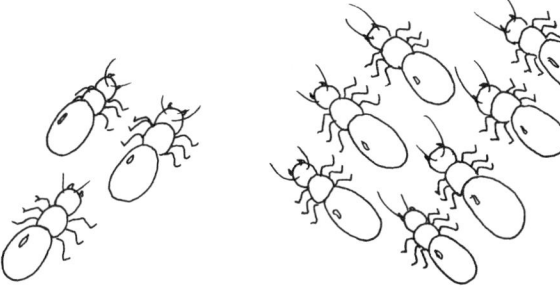

_____
- - - - -
_____

_____
- - - - -
_____

_____
- - - - -
_____

Circle Oh!

Find and 🖍 all the **circles** ○.

A Square Deal

Find and ⬭ all the **squares** ☐.

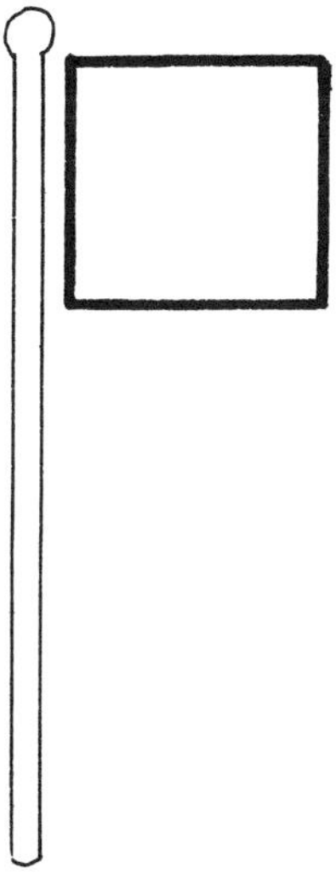

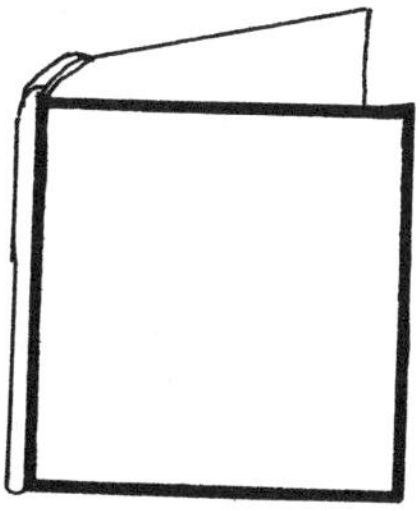

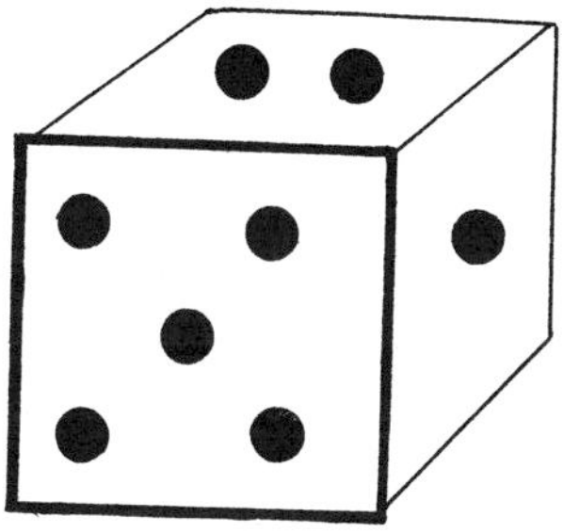

Ring the Triangle

Find and ⊂◯⊃ all the **triangles** △.

rectangle

The Rec Room

Find and ✏ all the **rectangles** ☐.

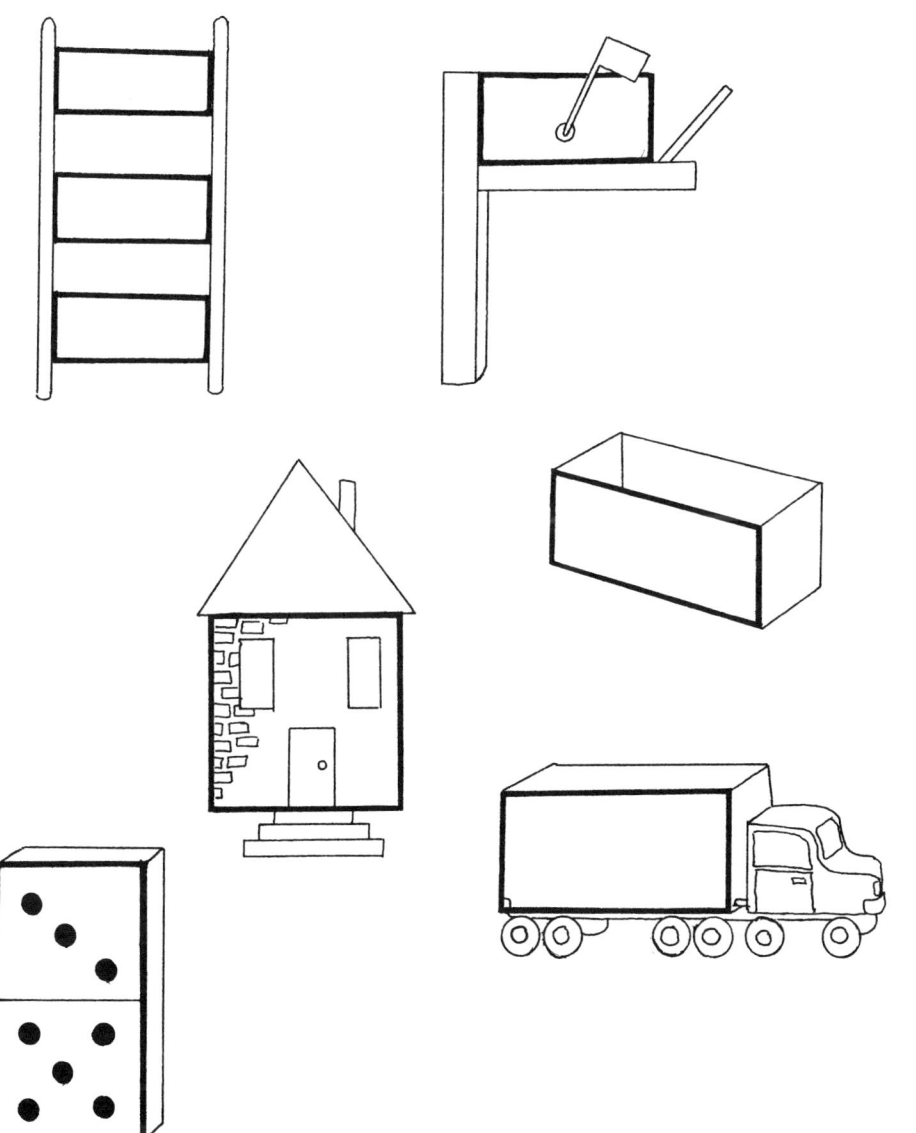

Ogling Ovals

Find and  all the **ovals** ◯.

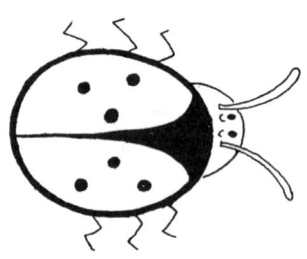

diamond

Diamonds Are Forever

Find and 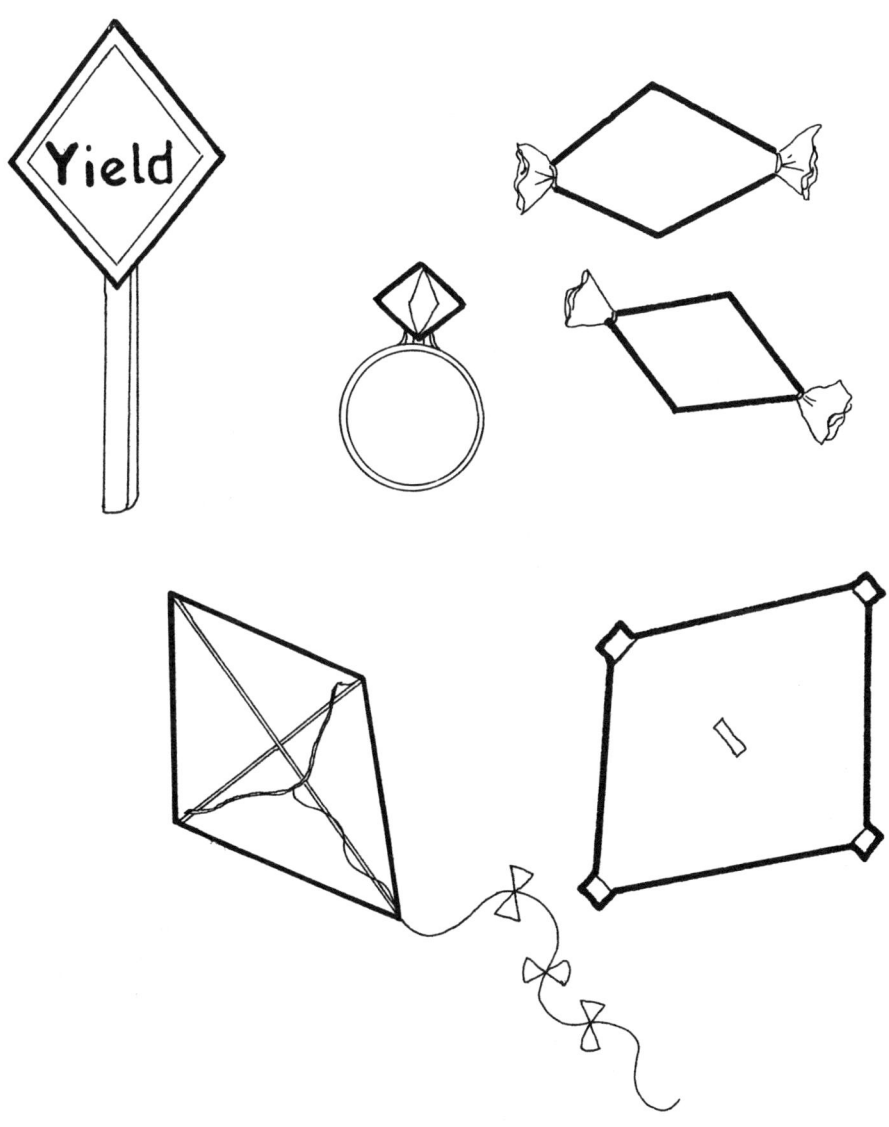 all the **diamonds** ◊.

Heart Throb

Find and 🖍 all the **hearts**

star

Starry Night

Find and 🖍 all the **stars** ☆.

What Day Is It?

✏️ in the missing numbers on the calendar. Answer the questions below.

Sunday	Monday	Tuesday	Wednesday	Thursday	Friday	Saturday
	1					
7				11	12	13
14	15	16	17	18	19	20
21	22	23	24	25	26	27
28	29	30	31			

How many Mondays do you see? _____

How many Fridays do you see? _____

How many Sundays do you see? _____

How many Wednesdays do you see? _____

calendars

✏️ a line to match the calendar with the correct season.

OCTOBER						
	1	2	3	4	5	6
7	8	9	10	11	12	13
14	15	16	17	18	19	20
21	22	23	24	25	26	27
28	29	30	31			

JANUARY						
		1	2	3	4	
5	6	7	8	9	10	11
12	13	14	15	16	17	18
19	20	21	22	23	24	25
26	27	28	29	30	31	

APRIL						
		1	2	3	4	5
6	7	8	9	10	11	12
13	14	15	16	17	18	19
20	21	22	23	24	25	26
27	28	29	30			

JULY						
		1	2	3	4	5
6	7	8	9	10	11	12
13	14	15	16	17	18	19
20	21	22	23	24	25	26
27	28	29	30	31		

summer

fall

spring

winter

What Time Is It?

 the time each clock shows.

_____ o'clock

_____ o'clock

_____ o'clock

_____ o'clock

time

✏️ _____ the time each clock shows.

🕐 = _____ o'clock

🕛 = _____ o'clock

🕑 = _____ o'clock

🕙 = _____ o'clock

🕛 = _____ o'clock

Happy Face

✏️ the missing numbers on the clock below.